I0817173

Nests

Julie Murray

Abdo Kids Junior
is an Imprint of Abdo Kids
abdobooks.com

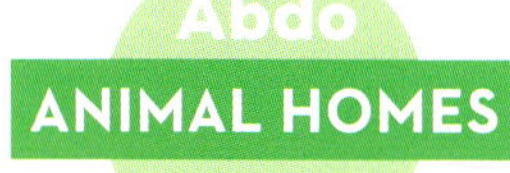

abdobooks.com

Published by Abdo Kids, a division of ABDO, P.O. Box 398166, Minneapolis, Minnesota 55439.

Printed in the United States of America, North Mankato, Minnesota.

052019

092019

Photo Credits: Alamy, iStock, Minden Pictures, Shutterstock

Production Contributors: Teddy Borth, Jennie Forsberg, Grace Hansen

Design Contributors: Christina Doffing, Candice Keimig, Dorothy Toth

Library of Congress Control Number: 2018963311

Publisher's Cataloging-in-Publication Data

Names: Murray, Julie, author.

Title: Nests / by Julie Murray.

Description: Minneapolis, Minnesota : Abdo Kids, 2020 | Series: Animal homes | Includes online resources and index.

Identifiers: ISBN 9781532185243 (lib. bdg.) | ISBN 9781644941218 (pbk.) | ISBN 9781532186226 (ebook) | ISBN 9781532186714 (Read-to-me ebook)

Subjects: LCSH: Animal housing--Juvenile literature. | Animal nests--Juvenile literature. | Nest building--Juvenile literature. | Animals--Habitations--Juvenile literature.

Classification: DDC 591.564--dc23

Table of Contents

Nests

Many kinds of animals live in nests.

Birds build nests. This is where they lay their eggs.

Many nests are made of **twigs**, mud, and grass. A robin's nest is shaped like a bowl.

A bald eagle's nest is big. It sits high in a tree. It is 6 feet (1.8 m) wide!

An ostrich's nest is on the ground. Its eggs are big!

Wasps live in a nest. Their nest grows and grows.

Ants live in a nest too. Their nest is **underground**.

Sea turtles dig a nest in the sand. They **bury** their eggs.

Have you seen a nest?

What Lives in a Nest?

alligators

gray squirrels

harvest mice

weavers

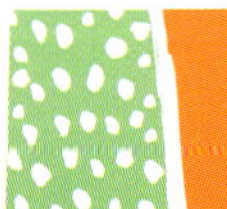

Glossary

bury
to cover in the ground with dirt or sand.

twig
a small, woody part of a tree that often grows from a tree's branch.

underground
a place beneath the earth's surface.

Index

Visit **abdokids.com** to access crafts, games, videos, and more!

Use Abdo Kids code

ANK5243

or scan this QR code!